A Theory of War in the Cyber Domain:

An Historical Perspective

David Aucsmith

Tera Nova Books, LLC

Tera Nova Books, LLC
Seattle, WA
www.teranovabooks.com

Book Cover:
Bonaparte crossing the St Bernard Pass, May 1800,
1801 painting by Jacques-Louis David (1748-1825), oil on canvas.

Library of Congress Control Number: 2023900083

ISBN-978-1-7349889-8-7

Printed in the United States of America

DEDICATION

To those who guard the gates

FORWARD

This monograph is a synthesis of two papers. The first paper, from which the title is derived, was originally written in 2012 when the notion of a fifth domain of warfare, the cyber domain, was just emerging, and militaries were struggling to develop a concept of multidomain operations that included the cyber domain. That paper has been required reading at various military academic institution.

The second paper was written in 2013 to extend some of the concepts of the original paper to address the nature of operations in cyberspace as irregular warfare.

This monograph is the result of a synthesis of those two papers into a coherent concept.

Table of Contents

Introduction

As technologies create new modes, or domains, of war, existing theories of war are, of necessity, reevaluated to adapt them to new capabilities of war fighting. The most obvious example of this was the development of theories of war after the introduction of heavier-than-air aircraft in the early 20th Century. Cyberspace, as a domain of war, is both new and pervasive. This paper addresses a theory of war in cyberspace from an historical perspective and describes how it has become necessary to change currently understood theories to make them applicable to the unique characteristics of cyberspace. It is about war both *from* and *within* the cyber domain. War *from* cyberspace is distinctly Western while war *within* cyberspace is distinctly not so.

In 1998, Admiral Arthur Cebrowski proposed the concept of *Network-Centric Warfare* as the basis for a theory of war *from* cyberspace.[1] That is, Cebrowski formulated a theory for using the capabilities of cyberspace to fight wars in the other domains. Once the capability exists to project force *from* the new domain *into* other domains, capabilities to wage war *within* the new domain soon follow. With the establishment of USCYBERCOM, we are now exploring how to wage war *within* the cyber domain.

Domains of War

Cyberspace is the newest of the five domains of war. William Lynn, United States Deputy Secretary of Defense, stated in 2010 that, "As a doctrinal matter, the Pentagon has formally recognized cyberspace as a new domain of warfare. Although cyberspace is a man-made domain, it has become just as critical to military operations as land, sea, air, and space. As such, the military must be able to defend and operate within

---

[1] Arthur K. Cebrowski, Vice Admiral, USN and John H. Garstka, "Network-Centric Warfare - Its Origin and Future," *U.S. Naval Institute Proceedings* 124, no. 1 (Jan 1998).

it."[2]

Technology Creates Domains of War

Land, as the original domain, was the sole domain of man's early warfare. It is still the principal domain of war, whether by sharp sticks or depleted uranium rounds. Technological inventions that change the mode of war and introduce new capabilities create domains in which control may be contested. The invention of ships created the maritime domain, airplanes the air domain, and satellites and missiles created the space domain. Now, computers and telecommunications have created the cyber domain.

Land is the Primary Domain

The inherent strength of land combat is that it carries the promise of achieving a decision.[3] As Sir Julian Corbett so eloquently put it when introducing a theory of war in the maritime domain, "Since men live upon the land and not upon the sea, great issues between nations at war have always been decided—except in the rarest cases—either by what your army can do against your enemy's territory and national life or else by the fear of what the fleet makes it possible for your army to do."[4] The same can be said of air, space, and cyber. Cyberspace, as a domain of war, is important primarily for what operations in cyberspace can enable forces in the other domains to do.

Domains Interact

Land is still the sole domain in which the outcome of conflicts is

---

[2] William J. Lynn, "Defending a New Domain," *Foreign Affairs* 89, no. 5 (Sep/Oct 2010): 97.

[3] Colin S. Gray, *Modern Strategy* (New York: Oxford University Press, USA, 1999), 214.

[4] Sir Julian Corbett, *Some Principles of Maritime Strategy.* (Annapolis, MD: United States Naval Inst., 1988), 54.

decided. It is the focus for which all strategic thought is relevant. The sea became relevant to war only when man could project power from the sea to the shore by either landing forces on shore or interdicting seaborne commerce essential for the survival of a people on land. Early amphibious operations, such as the landing at Marathon by the Persians in 490 BCE, illustrate the point regarding projecting force ashore. Likewise with commerce, as Alfred Thayer Mahan argued, in war, a nation that could protect its own maritime commerce while disrupting that of its opponent could shift the balance of national resources decisively in its favor.[5]

Once man could use the sea to affect the outcome of war on land, the sea became a domain of war in and of itself. Unique tactics, techniques, and procedures were developed to fight at sea, from the Greek trireme to the fully rigged ships carrying broadside artillery. The maritime domain became a strategic consideration not so much to win at sea as to achieve control of the sea to affect the outcome of conflicts on land. Navies did this through either destruction of the adversary's forces or interdiction of its commerce - the goal being to achieve command of the sea by being the master of maritime communications as defined by Corbett.[6] The ancient Athenians were masters of the sea in this respect and the Battle of Salamis is, of course, a prime example of that capability.[7]

Similarly, early aircraft sought to affect the war on land. Early theorists, such as Giulio Douhet, saw the airplane in observation, communications, ground support, and strategic bombing roles.[8] Once

---

[5] Alfred Thayer Mahan, *The Influence of Sea Power Upon History. 1660-1783. [With Maps and Plans.]* (Charleston: British Library, Historical Print Editions, 2011), 1-2.

[6] Corbett, 62.

[7] Herodotus, *The Landmark Herodotus: The Histories*, ed. Robert B. Strassler, trans. Andrea L. Purvis (New York: Pantheon, 2007), 637.

[8] Giulio Douhet, *The Command of the Air*, ed. Joseph Patrick Harahan and Richard H. Kohn (Tuscaloosa, AL.: University Alabama Press, 2009).

man could use the air to project power against the land, the air became a domain of war in and of itself. Unique tactics, techniques, and procedures were developed to fight in the air. The air domain became a strategic consideration, again, not so much to win in the air as to achieve control of the air to affect the outcome of conflicts on land (and sea), through either destruction of the opposition's forces (or their morale) or through interdiction of commerce. As Douhet says, "the command of the air is to prevent the enemy from flying, while assuring this freedom for oneself."[9]

Space followed much the same pattern. The early use of space was for observation and communications that affected the conduct of war on land. This capability has evolved to include, for example, early warning of the boost phase of intercontinental ballistic missiles, troop movements, and Global Positioning Systems. The space domain became a strategic consideration, like sea and air before it, not so much to win a war in space as to achieve control of the space domain to affect the outcome of conflicts on land (sea or air).

What then of cyberspace? Though computers and telecommunications have been a part of warfare from their very creation, it is with the emergence of network-centric warfare and related concepts that a capability to project power from cyberspace was developed that can affect the conduct of war on land (as well as sea, air, and space).[10] Thus, of necessity, cyberspace has become a domain of war in and of itself. We are developing unique tactics, techniques, and procedures to fight in cyberspace. Again, the strategic goal is not to win war fought in cyberspace *per se*; rather, it is to achieve control of cyberspace to affect the outcome of conflicts in the other domains by directly affecting the operation of military forces, disrupting commerce and degrading morale. General Keith Alexander, Commander of United States Cyber Command, stated before Congress,

---

[9] Ibid., 191.

[10] Cebrowski, 139.

> The cyber domain in some ways is like the air domain, in being a realm that had no relevance for military planning until all of the sudden a new technology offered access to it. A century ago the world's militaries had to learn to fight in the air, and they had to do so all at once in the midst of a world war. . . . The parallels with cyberspace seem obvious: freedom of action in cyberspace, like freedom of maneuver in the air, is crucial to the efficient employment of one's forces in all domains. Likewise, the loss of such freedom could impair the capabilities we have built in all the other domains.[11]

War *from* a Domain and War within a Domain

In the natural development of a domain of war, the capabilities of the new domain are first used to project power *from* the new domain onto another domain. For example, belligerents in World War I first used aircraft to observe and then later to attack the land domain. The enemy noted the new capability and developed a strategy for attacking the new capability from an existing domain. Shooting at planes with rifles or dedicated anti-aircraft artillery are examples of attacking the capabilities of the new domain (air) from an existing domain (land). At some point, as the capabilities in the new domain mature, strategies for countering the capabilities *within* the new domain are developed. In the air domain, fighter aircraft were developed to attack bombers and reconnaissance aircraft. The distinction between fighting *from* the domain and fighting *within* the domain is particularly important in the cyber domain. This paper will use the term "war *in* cyberspace" to denote both war *from* the domain and war *within* the domain when a

[11] Keith Alexander, General, USA, "Statement For the Record, Commander, United States Cyber Command," before the House Armed Services Committee, Washington, DC, 23 September 2010), http://www.defense.gov/home/features/2010/0410_cybersec/docs/USCC%20Command%20Posture%20Statement_HASC_22SEP10_FINAL%20_OMB%20Approved_.pdf (accessed December 30, 2011), 4.

distinction is not necessary.

War *from* the Cyber Domain

The creation of the cyber domain has followed the same steps as all previous domains. Computers and communications networks created the capability to exchange and manipulate information at great speed and on a large scale. Their existence became commonplace in the 1990's, as represented by the commercialization of the Internet. This capability created new business processes that leveraged information collection, processing, and distribution to improve the efficiency and profitability of companies. Businesses became information-based, network-centric, and Internet dependent.

In 1998, Admiral Arthur Cebrowski, then president of the Naval War College, proposed transforming the United States military by leveraging computer and communications capabilities for warfare, as business had leveraged them for commerce.[12] He proposed the concept of network-centric warfare (NCW) – a way to project power from cyberspace into the land, sea, air, and space domains.[13] As Cebrowski said, "Network-centric warfare enables a shift from attrition-style warfare to a much faster and more effective warfighting style characterized by the new concepts of speed of command and self-synchronization."[14] It enables "the wholesale and near instantaneous sharing of information within and among all elements of US and allied armed forces, irrespective of their locality or mode of operation."[15] Carl von Clausewitz wrote, "War is the realm of uncertainty; three quarters of the factors on which action is based are wrapped in a fog of greater

---

[12] Cebrowski, 139.

[13] Department of Defense, Office of Force Transformation, *The Implementation of Network-Centric Warfare* (Washington, DC: Department of Defense, 2005).

[14] Cebrowski, 148.

[15] Brice F. Harris, *America, Technology and Strategic Culture: a Clausewitzian Assessment* (New York: Routledge, 2009), 7.

or lesser uncertainty."[16] The promise of NCW is, as Admiral Bill Owens argues, that it lifts the fog of war.[17]

NCW envisions a "system of systems" that allows us to "sense" the battlefield with near perfect clarity, understand it, and strike with virtual impunity.[18] However, it is a strike *from* cyberspace. The strike is "sensed," planned, and directed from cyberspace, but the attack occurs in the physical world using weapons resident in the other domains of war that are *uniquely enabled by cyber capabilities.*

As NCW matured, a set of governing principles emerged that guided organization, training, and operation of armed forces. These principles augment the "time-tested principles of war —mass, objective, offensive, security, economy of force, maneuver, unity of command, surprise, [and] simplicity," with the information processing and communications of the computer and telecommunications networks of the cyber domain.[19] The principles of NCW begin with the requirement to fight first for information superiority.[20] That is, a network-centric military force must set as its first operational objective the achievement of an asymmetric advantage in the cyber domain of war.[21] Information superiority means having a dramatically better awareness of the battlespace than that of the adversary.[22]

In this sense, the requirement for cyber domain superiority is similar to the call for sea domain superiority of Mahan and the air

---

16 Carl von Clausewitz, *On War*, 1St Edition ed., trans. Michael Howard and Peter Paret (Princeton, NJ: Princeton University Press, 1989), 101.

17 William A. Owens, *Lifting the Fog of War* (Baltimore: The Johns Hopkins University Press, 2001).

18 Thomas X. Hammes, *The Sling and the Stone: On War in the 21st Century* (Minneapolis, MN: Zenith Press, 2006), 190.

19 Robert Leonhard, *The Principles of War For the Information Age* (New York: Presidio Press, 2000), 51-94.

20 Harris, 57.

21 Ibid.

22 Cebrowski, 148.

superiority of Douhet. Mahan argued the need for command of the sea through naval superiority was a prerequisite for a nation's security.[23] Similarly, the argument for information superiority sounds very much like Douhet's statement about command of the air.[24] Both Mahan and Douhet based their arguments on the realization that one's adversary would likely possess the same technology and could similarly use the domain to project power. The only real way to stop them was to have superiority in or control of the domain. The logical means of creating superiority in a domain of war is to fight *within* that domain for that control.

War *within* the Cyber Domain

Conflict within cyberspace is as old as the domain itself. "Hacking," the cyber equivalent of vandalism, occurred as soon as computers communicated over common networks. Attacks *within* the cyber domain manipulate information. Cyber weapons, such as malicious computer code, malformed input data, or non-standard protocols, can cause the underlying computer and communications systems to behave in ways that are either unintended or unexpected.

This sort of activity is distinct from actions *from* cyberspace. Using networked sensors and stored geospatial information to dispatch a cyber-controlled unmanned aerial vehicle (UAV) to fire a weapon at an electrical power generator and destroy it is an attack *from* cyberspace (through the air domain). Using the network to issue incompatible instructions to the digital controller for an electrical power generator and destroy it is an attack *within* cyberspace.[25]

---

[23] Jon Tetsuro Sumida, *Inventing Grand Strategy and Teaching Command: The Classic Works of Alfred Thayer Mahan Reconsidered (Woodrow Wilson Center Press)* (Charleston: The Johns Hopkins University Press, 1999), 44.

[24] Douhet, 24.

[25] In 2006, engineers at Idaho National Labs showed how a cyber-attack could cause any spinning machine connected to the power grid -- such as a generator, pump or turbine -- to self-destruct.

In 1983, the Department of Defense developed a set of fundamental principles for the security of computer and network systems.[26] They are:

Confidentiality – Preventing unauthorized disclosure of information.
Integrity – Preventing unauthorized modification of information.
Availability – Assurance that information is available when required.

Although the types of cyber-attacks may differ, all attacks within the cyber domain must subvert one or more of these principles to succeed. It is helpful to divide cyber-attacks into three different types, based on their objective and the United States legal authorities that apply.

War – (US Title 10) Attacks to deceive, deny, disrupt, degrade or destroy.
Espionage – (US Title 50) Spying by a government to discover military and political secrets.
Crime – (US Title 18) Theft, fraud, or other criminal acts.

One of the difficulties in defending against cyber-attacks is that the tools, techniques, and procedures used to attack are the same regardless of the type of attack. They differ only in their objective. This paper will address only those attacks with the objective of war.

The embodiment of war *within* the cyber domain is a non-violent attack against the information infrastructure of an enemy state or society by the subversion of one or more of the principles of security.[27] As it is non-violent in nature, is war in cyberspace *legally* war?[28] General

[26] Department of Defense, *Trusted Computer Security Evaluation Criteria - DOD 5200.28-Std* (Washington, DC: Department of Defense, 1983).

[27] Gray, 249.

[28] Leslie C Green, *Contemporary Law of Armed Conflict (Melland Schill Studies)*,

Alexander's statement before Congress leaves no doubt that in the United States, it is regarded as so, "the U.S. has affirmed that the International Law of Armed Conflict, which we apply to the prosecution of kinetic warfare, will also apply to actions in cyberspace."[29]

Peculiar Characteristics of Domains

The technology that is the genesis of a new domain of war imbues that domain with peculiar characteristics that are unique to the technology. When ships created the maritime domain, the ships brought uniqueness in, for example, speed and strategic mobility, which were unlike that of the land domain. Corbett notes this when he says that a characteristic of the maritime environment, "is the peculiar freedom and secrecy of movements at sea. As the sea knows no roads to limit or indicate our own lines of operation, so it tells little about those of the enemy."[30]

Similarly, airplanes brought speed and strategic mobility to the air domain completely unlike that of either the land or maritime domain. As Douhet noted, "The airplane has complete freedom of action and direction; it can fly to and from any point on the compass in the shortest time – in a straight line – by any route deemed expedient."[31] The airplane operates in, as Douhet called it, the 3rd dimension, unencumbered by terrain.[32]

Again, space has peculiar characteristics, mostly defined by physics. Space begins where the laws of orbital mechanics, rather than the laws of aerodynamics, govern the movement of vehicles.[33] It is both

---

3rd Revised edition ed. (Manchester: Manchester University Press, 2008), 387-91.

[29] Alexander, 7.

[30] Corbett, 134.

[31] Douhet, 9.

[32] Ibid.

[33] Arnold H. Streland, "Clausewitz on Space: Developing Military Space Theory

the freedom and the constraint of movement that are the unique characteristic of space as a domain: freedom, in that space vehicles can travel above – and "see" – any part of the land or sea; and constraint, in that travel is governed by the laws of orbital mechanics.

Seen in this light, one can propose that terrain is a peculiar characteristic of the land domain. Its significance is obvious in the theories of both Clausewitz and Jomini, and clearly not applicable to the other four domains. Clausewitz claims, when describing the genius of war, that, "a great peculiarity is given to the effect of this connection of War with country and ground."[34] He saw the ability to conceptualize war within the geography at hand as, perhaps, the greatest attribute of the genius of war.

The Peculiar Characteristics of the Cyber Domain

It is the domain's unique, peculiar characteristics, which are the consequences of the technology that created the domain, that challenge the assumptions inherent in existing theories of war. The nature of the cyber domain is unique in many ways. These unique, peculiar characteristics are collected together into six themes to facilitate analysis: violence, speed, communications, mobility, friction, and irregular war.

War as Violence

Is conflict in cyberspace war in the Clausewitzian sense? Clausewitz defined war as, "an act of force to compel our enemy to do our will."[35] By force, Clausewitz meant physical force. He says, "Force - that is physical force, for moral force has no existence save as expressed in the State and the law - is thus the *means* of war; to impose our will on

through a Comparative Analysis" (master's thesis, Air Command and Staff College, Air University, 1999), 12.

[34] Clausewitz, 109.

[35] Ibid., 75.

the enemy is its *object*."[36] By physical force, Clausewitz meant violence. He said, "War is a clash between major interests, which is resolved by bloodshed."[37] While war *from* cyberspace is ultimately violent, such as controlling a UAV to attack a land target, war *within* cyberspace may not be so. The very nature of the cyber domain is different and force in the cyber domain is a different concept.

Curiously, the *Department of Defense Dictionary of Military and Associated Terms*, defines neither "war" nor "warfare," although many variations of the latter (including electronic warfare, nuclear warfare, and unconditional warfare) are present.[38] The basic doctrinal publications of each of the US military services indicate that America's armed forces conceive of war in the tradition of Clausewitz.[39] To accommodate cyberspace, we may need to redefine war. Clausewitz's definition of war as "an act of force to compel our enemy to do our will" is still valid. However, we must expand the definition of *force* to include non-physical force. *Force* in this sense is a coercive act that either causes or threatens harm.

Force *within* cyberspace is therefore akin to the power to hurt. That is, it is coercive and destructive, but it does not imply the ability to physically seize and occupy. Thomas Schelling, in writing about nuclear deterrence, notes the important differences when he says, "This distinction between the power to hurt and the power to seize or hold forcibly is important in modern war, both big war and little war, hypothetical war and real war."[40] War *from* cyberspace can seize and forcibly hold. War *within* cyberspace can only hurt. Thus, war *within*

---

[36] Ibid., 75.

[37] Ibid., 149.

[38] *Department of Defense Dictionary of Military and Associated Terms: Joint Publication 1-02 (as Amended through 15 May 2011)* (Washington, DC: U.S. Government Printing Office, 2011).

[39] Harris, 102.

[40] Thomas C. Schelling, *Arms and Influence* (1966; repr., New Haven, CT.: Yale University Press, 2008), 6.

cyberspace is most effective for what it is *capable* of doing. Its greatest value is deterrence by the threat of potential harm.

Even if war starts solely *within* the cyber domain, war is a contest between intelligent actors, either of whom may introduce violence if they feel at a disadvantage.[41]

Tempo of Operations and Speed of Attack

Cyberspace is unique among domains of war in the tempo of operations and the speed of attack. Operations within cyberspace occur literally at the speed of light (minus network latency). Time, timing, and tempo are components of warfare in all domains, but the cyber domain is temporal in its essence. War in cyberspace is war in the 4th dimension (time).

Between 1977 and 1992, Colonel John Boyd developed the "OODA loop" (Observe, Orient, Decide, and Act) theory of warfare. Boyd's theory of conflict is well suited for analyzing war in the cyber domain, as it is more temporal than physical and spatial.[42] It focuses on the uncertainty created in an adversary who is significantly slower at processing situational information. William Lind summed up Boyd's theory as,

> Conflict can be seen as time-competitive observation-orientated decision-action cycles. Each party to a conflict begins by observing. He observes himself, his physical surrounding and his enemy. On the basis of his observation, he orients, that is to say, he makes a mental image or "snapshot" of his situation. On the basis of this orientation, he makes a decision. He puts the decision into effect, i.e., he acts. Then because he assumes his action has changed the

---

[41] David Lonsdale, "Clausewitz and Information Warfare," in *Clausewitz in the Twenty-First Century*, ed. Hew Strachan and Andreas Herberg-Rothe (Oxford: Oxford University Press, USA, 2007), 246.

[42] Robert B. Polk, "A Critique of the Boyd Theory - Is It Relevant to the Army?" *Defense Analysis* 16, no. 3 (2000): 258.

> situation, he observes again, and starts the process anew . . . With each action, the slower party's action is inappropriate by a larger time margin. Even though he desperately strives to do something that will work, each action is less useful than its predecessor; he falls farther and farther behind. Ultimately, he ceases to be effective.[43]

Before using Boyd's OODA loop to look at cyber war, it is necessary to look at how the tempo of operations has increased over time, while man's cognitive abilities have remained fixed. When Clausewitz and Jomini wrote in the early days of the industrial revolution, the speed of a man on horseback or on foot was the upper bound on the tempo of operations. This maximum tempo is an implicit assumption in their theories. The speed at which an attacker could close the distance to engage a defender meant that the attacker would be under observation by the defender for the time necessary for the defender to react - essentially eliminating tactical surprise. Jomini said, "Surprise of an army is now next to an impossibility."[44] The ability of the defender to observe, orient, decide, and act contributes to Clausewitz's declaration that defense is the stronger form of warfare, as the defense, through interior lines, always had time to mass at the point of attack. Clausewitz believed that an army could rarely obtain surprise in large actions. He said, "It therefore rarely happens that one State surprises another by a War, or by the direction which it gives the mass of its forces."[45]

Mechanization changed the tempo of warfare. Internal combustion engines became the upper bound of the speed of movement. Theorists such as Heinz Guderian modified doctrine to take advantage of mobility and shift the balance a little more toward offense. Guderian

---

[43] William S. Lind, *Maneuver Warfare Handbook* (Boulder, Colo.: Westview Press, 1985), 5-6.

[44] Baron Antoine-Henri de Jomini, *The Art of War* (Princeton, N.J.: Bottom of the Hill Publishing, 2011), 210.

[45] Clausewitz, 199.

saw that the speed and mobility of the tank could give an army the ability to surprise an enemy with an attack *en masse*.[46] The importance of speed and mobility was that they increased the tempo of operations. In terms of Boyd's OODA loop concept, speed and mobility decreased the time that was available for a defender to observe, orient, decide, and act – particularly the time for observation and orientation.

It was the airplane and the creation of air as a domain of war that caused the tempo of operations to become a critical component of strategy. Douhet and the early airpower theorists anticipated this. Douhet saw the airplane as the ultimate offensive weapon, "air power is a weapon superlatively adapted to offensive operations, because it strikes suddenly and gives the enemy no time to parry the blow."[47]

Attacks *within* cyberspace are nearly instantaneous, as is the command and control for attacks *from* cyberspace. There is no opportunity to observe the intentions of the offense, orient, and decide on a defense. The very nature of war in the cyber domain traps the defense within Boyd's OODA loop. In Boyd's theory of warfare, the offense aims to render the defense powerless by denying them the time to cope mentally with the rapidly unfolding—and naturally uncertain—circumstances of war.[48] In war *within* cyberspace, offense can always operate at a faster tempo or rhythm than defense. Thus, in cyberspace, offense is the stronger form of war.

Because of the near instantaneousness of cyber actions, preemptive attack becomes particularly appealing. This is an analogous situation to that of preemptive nuclear attack. As Schelling said,

> If the weapons can act instantaneously by the flip of a switch, a "go"

---

[46] Heinz Guderian, *Achtung-Panzer!: the Development of Tank Warfare* (London: Cassell, 1999), 205.

[47] Douhet, 18.

[48] David S. Fadok, "John Boyd and John Warden: Airpower's Quest for Strategic Paralysis," in *The Paths of Heaven: The Evolution of Airpower Theory*, ed. Phillip S. Meilinger (Maxwell Air Force Base, AL: Air Univ. Press, 1997), 364.

> signal, and can arrive virtually without warning to do decisive damage, the outcome of the crisis depends simply on who first finds the suspense unbearable.[49]

The offense in cyber war can assert both strategic and tactical surprise. If it can cause decisive damage, this creates a situation of great risk. Again, as noted by Schelling, when the premium is on haste and the advantage is to strike first, there is a great danger that peace will explode into all-out war.[50]

The situation in the cyber domain is the same as that initially envisioned by the air power theorists for the air domain, where defense is impossible. Ultimately, technological innovations, such as radar, established defense in the air domain. Perhaps technological innovations will similarly establish defense in the cyber domain. Until that happens, we are limited to hardening our systems to make them as resilient as possible, but due to friction (addressed later), they will never be completely impenetrable.

The only way to escape Boyd's trap is to know what your adversary will do *a priori*. If it is possible to know your adversary's intentions, tactics, and weapons, then it would be possible to progress through the observe, orient, and decide sections of Boyd's loop prior to attack, which would move the advantage back to the defender. It is possible to sense your adversary's intentions in cyberspace, but it depends on effective intelligence collection and analysis.[51] In short, one must have a presence in the cyber domain of the adversary.

Lines of Communications

Cyberspace exists solely as lines of communications. It has no other expression. Through cyberspace travel the command, control,

---

[49] Schelling, 225.

[50] Ibid., 227.

[51] David Aucsmith, "Rethinking Cyber Defense," *High Frontier* 7, no. 3 (May 2011): 36.

communications, intelligence, surveillance, and reconnaissance of armies and the essential trade and commerce of nations. Cyberspace is the global "center of gravity" for all aspects of national power, spanning the economic, technological, diplomatic, and military capabilities a country might possess.[52] Cyberspace has become, as Clausewitz said regarding a center of gravity, "the hub of all power and movement on which everything depends."[53]

Mahan described the sea as a "great highway" passing in all directions, a domain of trade and international communication with tremendous social and political importance.[54] He observed that certain lanes of travel, seaports, and communication routes will inevitably be preferred over others, and become recognized as "trade routes," which over time can have a significant influence on national and international commerce, and the course of history.[55] Cyberspace is the "great highway" of our age. It too has "ports" (e.g., data stores, sensors, weapons, and commercial systems) and "trade routes" (e.g., fiber optic cable and satellite links).

Corbett saw lines of communications as the vital component of the command of the sea. He said, "Command of the sea, therefore, means nothing but the control of maritime communications, whether for commercial or military purposes. The object of naval warfare is the control of communications, and not, as in land warfare, the conquest of territory."[56] In this respect, the cyber domain is similar to the sea domain. One cannot occupy the cyber domain, but one can control communications, both military and commercial. Command of cyberspace is the ability to enable one's own communications while

---

[52] Steve McPherson and Glenn Zimmerman, "Cyberspace Control," in *Securing Freedom in the Global Commons*, ed. Scott Jasper (Stanford, CA: Stanford Security Studies, 2010), 39.

[53] Clausewitz, 84.

[54] Mahan, 25.

[55] Ibid.

[56] Corbett, 94.

denying the enemy theirs.

Certainly, the communications of the cyber domain can be interdicted from other domains (such as an aircraft bombing a telecommunications facility), but they can also be interdicted by attacks within the cyber domain. Attacks against cyber lines of communications can destroy both the ability to project power *from* the cyber domain and to fight *within* the cyber domain. In addition to reducing military capacity, attacks against or within the cyber domain may also interdict commerce.

Like the sea of Mahan and Corbett's day, cyberspace is a conduit of commerce. As Corbett said of the sea, "Interference with the enemy's trade has two aspects. It is not only a means of exerting the secondary economic pressure; it is also a primary means towards overthrowing the enemy's power of resistance. Wars are not decided exclusively by military and naval force. Finance is scarcely less important."[57]

Physically attacking the cyber domain, that is, the destruction of the computers and network infrastructure is problematic because the cyber domain is a shared space, a global common like the sea. Corbett recognized the same issue with the sea domain. He said, "Since maritime communications are common, we as a rule cannot attack those of the enemy without defending our own. In military operations, the converse is the rule. Normally, an attack on our enemy's communications tends to expose their own."[58] This implies that attacks against cyberspace should be from *within* the cyber domain rather than *from* any other domain, unless the attacks *from* a domain can be precise enough to keep friendly lines of communications intact.

The implications of cyberspace being simply lines of communications are far reaching. A model similar to "a fleet in being" may apply. A fleet in being is a naval force that exerts influence on the adversary without ever leaving port. If the fleet were to leave port and

[57] Ibid., 102.

[58] Ibid., 100.

face the enemy, it might well lose in battle and no longer influence the enemy's actions, but while it remains safely in port the enemy must continually deploy forces to guard against it. A "cyber-force in being" would have the capability to disrupt commerce without "leaving port," so to speak. If a cyber-force were to act (leaves port), but fail to achieve a meaningful outcome, it would no longer have any influence on the adversary and could be safely ignored. Thus, its value is derived from the uncertainty of its effects.

The threat of commerce and military disruption from cyber-attack can act as a deterrent to others doing the same. This, however, assumes that the enemy is as vulnerable to cyber-attacks as you are. The less sophisticated the adversary, the less likely the deterrence value of a cyber-force in being.

In the sea domain, war has necessitated the military protection of commerce. Maritime theory acknowledges the tension between protecting commerce and destroying the enemy.[59] Cyberspace doctrine has explicitly rejected any military responsibility to protect commerce by drawing a distinction between the military responsibility for the *.gov* and *.mil* domains and the civil (Department of Homeland Security) responsibility for the *.com* domain.[60] This distinction is not consistent with the historic perspective as it assumes that commerce resides solely with the legal jurisdiction of the United States and not through global lines of communications. It is akin to only protecting costal trade in the maritime domain.

Strategic Mobility

The cyber domain compresses both time and space. Actions in cyberspace are near instantaneous without regard to geography. One can attack anywhere from anywhere. Like strategic bombing, cyber-attacks could bypass enemy forces to strike directly at the perceived

[59] Ibid., 133.

[60] Alexander, 8.

center of gravity.[61] It is the quintessential domain for strategic mobility.

Omnipotent strategic mobility creates an asymmetry between offense and defense. Offense is free to attack anywhere and defense must defend everywhere. This naturally leads to a quotation from Frederick the Great, "He who defends everything, defends nothing." Defense in cyberspace is inferior to offense.

There are historical analogues to this type of asymmetry. Certainly, the early airpower theorists envisioned a similar asymmetry for the airplane, but there is also the analogy of the submarine. The initial response to the threat of both airplane and submarine was to attack them at their bases – the one place where they were known to be. Later, technological innovations, such as radar and sonar, reduced their strategic mobility.

Our current strategy for cyber defense is to defend only what is important by hardening our systems (building cyber "bunkers") and attacking systems that cyber-attack us (attack their bunkers). It is possible that technological innovation will provide a counter to strategic mobility in cyberspace, but we have yet to see it.

Friction

Friction, as uncertainty in war, was one of Clausewitz's greatest contributions. Clausewitz said that, "Friction is the only concept that more or less corresponds to the factors that distinguish real war from war on paper."[62] Friction is the sum of all unknowns and chance that accumulate in any large, complicated endeavor such as war. Tasks easily done at small scale become difficult at large scale. It causes reality not to conform to theory.

The proponents of war in cyberspace seldom acknowledge that cyberspace is replete with complexity, chaos, and friction. War in cyberspace depends on a vast collection of sensors, data stores,

---

[61] Lonsdale, 237.

[62] Clausewitz, 119.

processing algorithms, operators, and weapons - all of this networked together across a global collection of communication nodes, many of whose exact architecture and function are unknown and unverifiable. Both war *from* and war *within* cyberspace depend on this technological web to provide "situational awareness," and both have limitations – but for completely opposite reasons.

The promise of war *from* cyberspace, NCW, is that it will lift the fog of war. It will be able to accomplish this because of the vast sensor and information web to which it has access. However, as has been noted by cognitive systems researchers, "The belief that more data or information automatically leads to better decisions is probably one of the most unfortunate mistakes of the information society."[63] War *from* cyberspace results in far too much information and far too little intelligence. The synthesis of information into intelligence remains a manual problem and does not scale linearly. Like much of Clausewitz's friction, situational awareness at small scale becomes difficult at large scale.

War *within* cyberspace suffers from the opposite problem. Owing to the attacker's strategic mobility (discussed previously), the defender has no awareness – no information - of attacks or attackers until they strike – if then. All operations take place in a fog of war. The only way to achieve situational awareness *within* cyberspace is to sense beyond one's own networks. That is, one must have a presence in the cyber domain of the adversary.[64] Even automated defense requires some capability to sense the attack.

Clausewitz used the term "friction" to describe the sum of uncertainty because it was a concept emerging from the science of his day. He was in the midst of a scientific revolution centered on Newtonian physics.[65] Today, we, too, are in the midst of a scientific

---

[63] David D. Woods and Erik Hollnagel, *Joint Cognitive Systems: Patterns in Cognitive Systems Engineering* (Boca Raton: CRC Press, 2006), 7.

[64] Aucsmith, 36.

[65] Thomas S. Kuhn, *The Structure of Scientific Revolutions,* [2d ed., vol. 2, of

revolution. Ours is centered on nonlinear systems. Research into nonlinear systems, specifically chaos and complexity, can give us additional insight into Clausewitz's friction.[66]

Cyberspace is nonlinear. Outputs are not proportional to inputs. Systems in cyberspace exhibit "organized complexity" and are highly interdependent.[67] They process information concurrently and cannot be reduced to state diagrams. As Alan Beyerchen notes, "Over time they change in ways that affect how and even which variables interact. The exact behavior of these systems is nearly impossible to predict, because they adapt to changing conditions."[68] In short, the exact behavior of cyberspace to a set of events is unknown *a priori* and may well be different each time.

Nonlinearity adds friction to war in the cyber domain through three mechanisms: complexity, chaos, and speed. Complexity, as noted above, increases with connectivity, causing a decrease in competency and an increase in fragility. Chaos is a state of nonlinear systems when they are highly sensitive to initial starting conditions – which, as in war, cannot be accurately known. Chaotic systems, like the weather, have an upper bound on what it may be *possible* to know about the system's behavior. Lastly, the speed and tempo of operations in cyberspace, addressed earlier, create uncertainty. Returning to Boyd's OODA loop, one normally thinks of cycling faster than an adversary can respond. In cyberspace, it is possible to cycle faster than the originator can respond. If your response completes before you have an ability to observe and orient from the outcome of your own act, yet you continue to act, you

---

International Encyclopedia of Unified Science. Foundations of the Unity of Science, V. 2, No. 2 (Chicago: University of Chicago Press, 1970), 67-9.

[66] Alan Beyerchen, "Clausewitz, Nonlinearity, and the Unpredictability of War," *International Security* 17, no. 3 (Winter 1992-1993): 59-90.

[67] Alan Beyerchen, "Clausewitz and the Non-Linear Nature of Warfare: Systems of Organized Complexity," in *Clausewitz in the Twenty-First Century*, ed. Hew Strachan and Andreas Herberg-Rothe (Oxford: Oxford University Press, 2007), 53-54.

[68] Ibid.

are stuck in a self-defeating OODA loop.

What is more important in understanding war in cyberspace is not how much we can know, but how much we cannot know with mathematical certainty. It is the limit on certainty, defined by nonlinearity and expressed in cyber war as Clausewitzian friction, which is important. The science of nonlinearity has closed the door forever on our ability to remove friction through the application of technology.[69]

Western War

War *from* the cyber domain, as this paper argues, is an embodiment of Western Warfare. Irrespective of a continuity of a Western Way of War from the Greek polis to contemporary times, there is a distinct way Western states wage war, though defining it is somewhat difficult. It is the warfare of Clausewitz, Jomini, Schlieffen, Mahan, Corbett, Douhet, Guderian, Schwarzkopf, Cebrowski, and many others. It is, as Geoffrey Parker notes, reliant on superior technology used by highly disciplined troops focused on the destruction – the annihilation - of the enemy's forces.[70] Others, such as Victor Davis Hanson, and Russell Weigley, have expanded on, or modified, Parker's theme, but the essence has remained the same. As Hanson notes, Western War is lethal and it is so lethal because it is amoral – not tied to anything but military necessity.[71]

War *from* the cyber domain is Western War, but it is more. It is Western War fought with the peculiar characteristics of the technology of cyberspace. It emphasizes speed, jointness, knowledge, and precision, as championed by the Office of Force Transformation and

---

[69] Barry D. Watts, *Clausewitzian Friction and Future War (McNair Paper 52),* (Washington, DC: National Defense University, 2004), 108.

[70] Geoffrey Parker, ed., *The Cambridge History of Warfare* (New York: Cambridge University Press, 2005), 1-11.

[71] Victor Hanson, *Carnage and Culture: Landmark Battles in the Rise to Western Power* (New York: Anchor, 2002), 21.

the Office of the Secretary of Defense.[72] A precise definition of Western War is not required in the context of cyberspace as the goal of this paper is to contrast it with non-Western War. In this context, Western War is aggressive, technologically enabled, lethal war that aims to defeat an enemy's forces as quickly as possible through superior battlefield knowledge and maneuver, while exposing its own combatants to minimal risk.

Non-Western War

Addressing war *within* the cyber domain as non-Western War is much like describing the animal kingdom as non-elephants. In this context, it is the war of Sun Tzu, T. E. Lawrence, Mao Tse-Tung, Che Guevara, Carlos Marighella, Hamas, and others. While Western War stresses a direct confrontation, a decisive battle, to destroy the enemy, non-Western War, as Sun Tzu wrote, stresses speed, surprise, and deception.[73]

Similarly, the *Handbook for Volunteers of the Irish Republican Army* notes, "In regular warfare the tactical objective is to destroy the enemy in battle by concentrating superior numbers at a decisive time and place. The guerrilla strikes not one large blow, but many small ones; he hits suddenly, gnaws at the enemy's strength, achieves surprise, disengages himself, withdraws, disperses and hits again."[74]

In this context, non-Western War is war of attrition, based on surprise, deception and maneuver rather than decisive engagements. It focuses on a large number of small battles rather than a small number of large battles. Non-Western War, in this context, is the indirect, circuitous approach advocated by Chinese *Shih*-strategy.[75]

---

[72] Department of Defense, 7-10.

[73] Sun-Tzu, "Sun-Tzu's Art of War," in *The Seven Military Classics of Ancient China*, trans. Ralph D. Sawyer (New York: Basic Books, 2007), 155.

[74] Anonymous, *Handbook for Volunteers of the Irish Republican Army: Notes on Guerrilla Warfare* (Boulder, CO: Paladin Press, 1996), 6.

[75] William H. Mott IV and Jae Chang Kim, *The Philosophy of Chinese Military*

Non-Western War comes in many different guises, but this paper will focus on guerilla warfare, first in general and then as it pertains to war in the cyber domain.

Guerrilla War

Samuel P. Huntington notes that, there is a distinction between *types of war* and *forms of war*.[76] Guerrilla warfare is a *form* of warfare and not a *type* of war. It may, for example, be used as part of a general war between nations or as part of a revolutionary struggle between a nongovernmental group and a government. Guerrilla warfare is, as Huntington defines it, "a form of warfare by which the strategically weaker side assumes the tactical offensive in selected forms, times, and places."[77]

It is the preferred form of war for those at a significant military disadvantage. Its practitioners claim that, "A small nation fighting for freedom can only hope to defeat an oppressor or occupying power by means of guerrilla warfare. The enemy's superiority in manpower, resources, material, and everything else that goes into waging war can only be overcome by the correct application of guerrilla methods."[78] Mao Tse-Tung, who was the first to systematically study and write about guerrilla warfare, said that it is a strategy that must be based primarily on alertness, mobility, and attack.[79]

It is distinguished, in part, from conventional, or Western, warfare in that Guerrilla warfare does not concentrate on a decisive battle,

---

Culture: Shih Vs. Li (New York: Palgrave Macmillan, 2006), 12.

[76] Samuel P. Huntington, "Guerrilla Warfare in Theory and Policy," in *Modern Guerrilla Warfare*, 1St Edition ed., ed. Franklin Mark Osanka (New York: The Free Press of Glencoe, 1962), xvi.

[77] Ibid.

[78] Anonymous, *Handbook for Volunteers of the Irish Republican Army: Notes on Guerrilla Warfare*, 6.

[79] Mao Tse-Tung, *On Guerrilla Warfare*, trans. Samuel B. Griffith II (Urbana: University of Illinois Press, 2000), 46.

rather, as Mao says, "In guerrilla warfare, select the tactic of seeming to come from the east and attacking from the west; avoid the solid; attack the hollow; attack; withdraw; deliver a lighting blow, seek a lighting decision. When guerrillas engage a stronger enemy, they withdraw when he advances; harass him when he stops; strike him when he is weary; pursue him when he withdraws."[80]

Guerrilla warfare is a very old form of warfare. During the Hundred Years War (1337 – 1453), the English lost most of their lands in France as a result of guerrilla war.[81] The efficiency of the longbow made the English superior on the battlefield so the French simply refused to fight them. Instead, French troops "raided them at night, ambushed their convoys, and harried their fortified camps and towns. [They] made no effort to drive them from France; [they] merely made it uncomfortable for them to stay!"[82] This is the essence of guerrilla warfare.

A Note on Hacking and Hactivism

It is important to note that "hacking," as it is generally understood, is not warfare and is not within the scope of this paper. It is largely criminal behavior and ranges from experimentation and vandalism to fraud and theft. If those who practice hacking do so at the behest of a government, they may be waging war – a criminal act committed by a government may be an act of war. As noted earlier, it depends on the objective of the attack.

Unconventional War as War *within* the Cyber Domain

As previously argued, war *from* cyberspace is network-centric

---

[80] Ibid.

[81] Frederick Wilkins, "Guerrilla Warfare in Theory and Policy," in *Guerrilla Warfare*, 1St Edition ed., ed. Franklin Mark Osanka (New York: The Free Press of Glencoe, 1962), 4.

[82] Ibid.

warfare. It "generates increased combat power by networking sensors, decision makers, and shooters to achieve shared awareness, increased speed of command, high tempo of operations, greater lethality, increased survivability, and a degree of self-synchronization."[83] It is technologically based war that aims to be short, sharp, and decisive – the epitome of Western war. The cost of entry is very high and the cost to sustain the capability even higher.

War *within* cyberspace is fundamentally different. As Secretary Lynn noted,

> Cyberwarfare is asymmetric. The low cost of computing devices means that U.S. adversaries do not have to build expensive weapons, such as stealth fighters or aircraft carriers, to pose a significant threat to U.S. military capabilities. A dozen determined computer programmers can, if they find a vulnerability to exploit, threaten the United States' global logistics network, steal its operational plans, blind its intelligence capabilities, or hinder its ability to deliver weapons on target.[84]

Adversaries are not limited to nation states. The low cost of entry allows individuals or groups to pursue war *within* cyberspace using unconventional warfare. More importantly, relatively minor actors can fight *within* the cyber domain and have global reach. Guerrilla war in cyberspace can negate the wealth and technological advantage of Western states.

War *within* cyberspace emphasizes many of the attributes of unconventional warfare. Specifically, it shares the characteristics of speed, mobility, and anonymity with that of guerrilla warfare. More importantly, it enlarges the scope and effect of guerrilla warfare. Hitherto, guerrilla war could only be fought successfully on one's home territory, as it was necessary to rely on the local population for

---

[83] Department of Defense, 4.

[84] Lynn.

support and anonymity. With cyberspace, it is now possible to project guerrilla war beyond one's home territory. It is now possible to gather supporters linked not by geography, but by ideology.

Guerrilla War in Cyberspace

War *within* cyberspace is essentially the low intensity, attack and retreat, anonymous warfare characteristic of classic guerrilla warfare fought by non-state actors. Guerrilla war in cyberspace, like the physical world, can be fought either in support of a conventional army, as in the case of Mao fighting the Japanese, or as an independent effort, such as Castro fighting the Batista regime. Che noted the two different uses when he said, "in current interpretations there are two different types of guerrilla warfare, one of which [is] – a struggle complementing great regular armies, " and "the other type, the case of an armed group engaged in struggle against the constituted power."[85]

Guerrilla War to Support Conventional Cyber War

Historically, guerrillas have mostly been used for independent war, not associated with a conventional army. Small bands of men have surprised and fought invading armies by attacking isolated groups of the enemy. As Frederick Wilkins notes, "Such men have found a place in the legends of many lands, but they made small impression in military history."[86] Mao noted that the purpose of guerrilla war was not to win in itself, but to augment regular forces. He notes that, "The concept that guerrilla warfare is an end in itself and that guerrilla activities can be divorced from those of the regular forces is incorrect."[87] "If we do not fit guerrilla operations into their

---

[85] Ernesto Che Guevara, *Guerrilla Warfare* (New York: bnpublishing.com, 2007), 10.

[86] Wilkins, 3.

[87] Mao, 55.

proper niche, we cannot promote them realistically."[88]

Mao said that the responsibility of guerrilla forces was "to harass and weaken large forces; to attack enemy lines of communications; to establish bases capable of supporting independent operations in the enemy's rear; to force the enemy to disperse his strength; and to coordinate all these activities with those of the regular armies on distant battle fronts."[89] Similarly, Clausewitz saw guerrillas as only useful in support of conventional forces. He said, "To be realistic, one must therefore think of a general insurrection within the framework of a war conducted by the regular army and coordinated in one all-encompassing plan."[90]

As war *within* the cyber domain can only harm or threaten harm, not seize and forcibly hold, the question of guerrilla war in cyberspace is how to achieve the greatest impact. As noted earlier, Western War is now fought as war *from* the cyber domain. As such, it is highly sensitive to attacks that deceive, deny, disrupt, degrade or destroy the infrastructure of cyberspace. The speed of cyber-attacks, the strategic mobility of cyberspace, and the anonymity of war fought *within* the cyber domain against lines of communications, make guerrilla war in cyberspace a particularly attractive adjunct to conventional warfare.

Cyber warfare, by United States defense doctrine, is to part of conventional warfare.[91] Russia used attacks *within* the cyber domain to disable and disrupt the military command and civil government of

---

[88] Ibid.

[89] Mao, 53.

[90] Clausewitz, 480.

[91] Keith Alexander, General, USA, "Statement For the Record, Commander, United States Cyber Command," before the House Armed Services Committee, Washington, DC, 23 September 2010), http://www.defense.gov/home/features/2010/0410_cybersec/docs/USCC%20Command%20Posture%20Statement_HASC_22SEP10_FINAL%20_OMB%20Approved_.pdf (accessed December 30, 2011), 4.

Georgia prior to their August 2008 invasion.[92] It was guerrilla war fought *within* the cyber domain in support of conventional forces and a national objective.

Guerrilla War as Independent Cyber War

Guerrilla war need not be fought in support of conventional forces. It may be fought as an independent war. Unlike Mao, this was the doctrine of Che Guevara. However, fighting an independent guerrilla war has several weaknesses. First, it normally requires an extended time to work. It is not a quick way to fight an adversary. This requires a long-term commitment from those perusing the fight. It is usually very difficult to maintain that level of commitment. Second, it requires local populist support t so that the guerrillas can remain hidden. When they fail to do so, as Che Guevara found out in Bolivia, the guerrilla is successfully attacked. Third, it can only be fought on one's "home ground." This is because of the need to maintain the support of the local population.

Independent guerrilla war *within* cyberspace overcomes two of these objections. Yes, it still requires a sustain level of commitment over an extended period. However, it does not require the support of a local populous. In fact, as described earlier, ideology that is independent of geography can now link guerrilla forces. Nor does it require "home" ground. Local support for resources and anonymity is no longer required. If the independent guerrilla can meet their objectives without the ability to seize and forcibly hold, guerrilla war *within* cyberspace would seem to be the ideal mode of war.

Urban Guerrilla Warfare as Cyber Warfare

Urban guerrilla war is a very old mode of warfare. Thucydides describes urban guerrilla warfare involving Theban troops, who were occupying the city of Plataea in 431 BCE,

---

[92] John Markoff.

> The Thebans, finding themselves outwitted, immediately closed up to repel all attacks made upon them. Twice or thrice, they beat back their assailants. But the men shouted and charged them, the women and slaves screamed and yelled from the houses and pelted them with stones and tiles; besides, it had been raining hard all night; and so at last their courage gave way, and they turned and fled through the city. Most of the fugitives were quite ignorant of the right ways out, and this, with the mud, and the darkness caused by the moon being in her last quarter, and the fact that their pursuers knew their way about and could easily stop their escape, proved fatal to many.[93]

What distinguishes urban guerrilla warfare from the rural, or regular, guerrilla warfare is the infrastructure available to the guerrilla. Rural warfare allows the guerrilla to seek isolated areas in which to establish bases and establish supply lines. It allows for mobility because movement will not be observed and anonymity because a rural area cannot be easily surrounded and searched.

Mao maintained that guerrillas needed at least one safe geographical base. That meant some place, some area within the boundaries of the state, that counterinsurgent forces could not effectively reach, or would not attempt, for whatever reasons, to occupy. Such a base would be invaluable to guerrillas: within it they could train and indoctrinate recruits; care for the sick and wounded; and stockpile munitions, food, and medicine.[94] Hence, for Mao protracted guerrilla warfare would not be possible in a small country or within the constricted area of a city.

Guerrilla warfare *within* cyberspace does not have, nor need, such support structures. Belligerents can be completely anonymous to

---

[93] *The Landmark Thucydides: A Comprehensive Guide to the Peloponnesian War*, ed. Robert B. Strassler (New York: Free Press, 1996), 91.

[94] Mao Tse-tung, "On the Protracted War," *Selected Works of Mao Tse-tung* (London: 1954), vol. 2, 171.

those around them. They need no base of operations. Cyberspace, in its virtual sense, is infinitely large and cannot be easily surrounded or searched. Counterinsurgent forces cannot effectively reach nor occupy the guerrilla's base of operations. Mao wrote, "The people are like water and the army is like fish."[95] By this, he meant the army hides and flourishes among the people and it is among the people (polity) that is the natural habitat of a revolutionary army. In cyberspace, fish swim in the masses of the cities.

In the physical world, urban guerrilla warfare has been a failure – precisely because it lacks the support infrastructure of rural guerrilla warfare. Nowhere in the twentieth century has urban guerrilla warfare achieved success. The record of urban guerrilla warfare is one of total and tragic defeat.[96] However, in the cyber domain, the support systems of the guerrilla are in the anonymity of the connected masses. War *within* the cyber domain more closely shares the attributes and needs of the urban guerrilla rather than those of the rural guerrilla.

Observations on War *from* Cyberspace

War *from* cyberspace is not a radical departure from modern military thought. It is an extension of Western war through another domain. The cyber domain offers a speed and tempo of operations not previously attainable and a level of intelligence that may thin (but not remove) the fog of war.[97] However, it also increases the brittleness of forces, the uncertainty of operations, and the friction of actions. More information and more technology will not ultimately bring about complete knowledge but it will allow us to leverage what certainty we do have over a greater geography and at higher speed.

---

[95] Ibid., 183.

[96] Anthony James Joes, *Urban Guerrilla Warfare* (Lexington: The University Press of Kentucky, 2007), 157.

[97] Admiral Bill Owens, conversation with author at the "Cyber Deterrence Workshop," 23 February 2010.

Observations on War *within* Cyberspace

War *within* the cyber domain is radically different from anything that has come before. It is non-violent war whose power to hurt may best be used as a "cyber force in being" to deter the actions of others – provided they have similar military and commercial dependencies. For those who do not, it is inherently asymmetrical allowing a modestly funded adversary great strategic mobility and the option to strike first. It does not allow quick, sharp, decisive warfare. It favors the cyber equivalent of urban guerrilla warfare where attackers may simply hide in the anonymity of the crowd.

War *within* cyberspace continues to evolve in both capability and complexity, creating complex systems that are simultaneously robust to large events and brittle to small ones. It is a global common, local to all, that compresses both space and time. It maintains an underlying uncertainty that cannot be engineered away, but may be exploited by an attacker.

Conclusions

Computers and communications have created cyberspace as a domain of war, a sphere of influence for which control may be contested. The technology that created cyberspace imbued it with unique capabilities. Unlike the other four domains of war, fighting *from* the domain and fighting *within* the domain have very different characteristics. War *from* the cyber domain was initiated with the formalization of network-centric warfare. It is an extension of Western war in a new domain. Western War is aggressive, technologically enabled, lethal war that aims to defeat an enemy's forces as quickly as possible through superior battlefield knowledge and maneuver, while exposing its own combatants to minimal risk. War *within* the cyber domain more closely resembles non-Western, unconventional warfare – specifically, guerrilla warfare. It uses the stealth, anonymity, speed, and mobility of cyberspace to attack the information assets of superior forces with near impunity.

Bibliography

Alexander, Keith, General, USA. "Statement For the Record, Commander, United States Cyber Command." Before the House Armed Services Committee, Washington, DC, 23 September 2010. http://www.defense.gov/home/features/2010/0410_cybersec/docs/USCC%20Command%20Posture%20Statement_HASC_22SEP10_FINAL%20_OMB%20Approved_.pdf (accessed December 30, 2011).

Allen, Patrick D., and Dennis P. Gilbert. "The Information Sphere Domain Increasing Understanding and Cooperation." NATO Cooperative Cyber Defence Centre of Excellence Tallinn, Estonia. http://www.ccdcoe.org/publications/virtualbattlefield/09_GILBERT%20InfoSphere.pdf (accessed January 9, 2012).

Arquilla, John, David Ronfeldt, and editors. *In Athena's Camp: Preparing For Conflict in the Information Age*. New York: Rand Corporation, 1997.

Aucsmith, David. "Rethinking Cyber Defense." *High Frontier* 7, no. 3 (May 2011): 35-37.

Barnett, Thomas P. M. "The Seven Deadly Sins of Network-Centric Warfare." *U.S. Naval Institute Proceedings* 125, no. 1 (January 1999): 36-9.

Berkowitz, Bruce. *The New Face of War: How War Will Be Fought in the 21st Century*. New York: Free Press, 2003.

Beyerchen, Alan. "Clausewitz, Nonlinearity, and the Unpredictability of War." *International Security* 17, no. 3 (Winter 1992-1993): 59-90.

Blaker, James R. *Transforming Military Force: the Legacy of Arthur Cebrowski and Network Centric Warfare*. Westport, CT.: Praeger, 2007.

Cebrowski, Arthur K., Vice Admiral, USN. "Network-Centric Warfare: An Emerging Response to the Information Age." *Military Technology* 27, no. 5 (May 2003): 16-22.

_______ and Thomas P. M. Barnett. "The American Way of War." *U.S. Naval Institute Proceedings* 129, no. 1 (January 2003): 42-44.

_______ and John H. Garstka. "Network-Centric Warfare - Its Origin and Future." *U.S. Naval Institute Proceedings* 124, no. 1 (Jan 1998): 139.

Clausewitz, Carl von. *On War*. 1St Edition ed. Translated by Michael Howard and Peter Paret. Princeton, NJ: Princeton University Press, 1989.

Corbett, Sir Julian. *Some Principles of Maritime Strategy*. Annapolis, MD: United States Naval Inst., 1988.

Creveld, Martin van. *The Transformation of War*. New York: Free Press, 1991.

Dahl, Erik J., Commander, USN. "Network-Centric Warfare and the Death of Operational Art." *Defence Studies* 2, no. 1 (Spring 2002): 1-24.

DeBlois, Bruce M., Richard L. Garwin, R. Scott Kemp, and Jeremy C. Marwell. "Space Weapons: Crossing the U.S. Rubicon." *International Security* 29, no. 2 (Autumn 2004): 50-84.

Department of Defense. *Department of Defense Dictionary of Military and Associated Terms: Joint Publication 1-02 (as Amended through 15 May 2011)*. Washington, DC: U.S. Government Printing Office, 2011.

_______. Office of Force Transformation. *The Implementation of Network-Centric Warfare*. Washington, DC: Department of Defense, 2005.

_______. *Trusted Computer Security Evaluation Criteria - DOD 5200.28-Std*. Washington, DC: Department of Defense, 1983.

_______. "USSTRATCOM Space Control and Space Surveillance." USSTRATCOM. http://www.stratcom.mil/USSTRATCOM_Space_Control_and_Space_Surveillance (accessed January 9, 2012).

Derian, James Der. "Virtuous War/virtual Theory." *International Affairs* 76, no. 4 (October 2000): 771-88.

Douhet, Giulio. *The Command of the Air*. Edited by Joseph Patrick Harahan and Richard H. Kohn. Tuscaloosa, AL.: University Alabama Press, 2009.

Echevarria, Autulio J. "Clausewitz's Center of Gravity: It's Not What We Thought." *Naval War College Review* (Winter 2003): 108-23.

Geers, Kenneth. "Cyber Weapons Convention." *Computer Law and Security Review* 26, no. 5 (September 2010): 547-51.

_______. "The Challenge of Cyber Attack Deterrence." *Computer Law and Security Review* 26, no. 3 (May 2010): 298-303.

Gompert, David C., Irving Lachow, and Justin Perkins. *Battle-Wise: Seeking Time-Information Superiority in Networked Warfare.* Washington, D.C.: National Defense University, 2006.

Gray, Colin S. *Modern Strategy*. New York: Oxford University Press, USA, 1999.

Green, Leslie C. *Contemporary Law of Armed Conflict (Melland Schill Studies)*. 3rd Revised edition ed. Manchester: Manchester University Press, 2008.

Guderian, Heinz. *Achtung-Panzer!: the Development of Tank Warfare.* London: Cassell, 1999.

Guha, Manabrata. *Reimagining War in the 21st Century: from Clausewitz to Network-Centric Warfare*. New York: Routledge, 2011.

Hammes, Thomas X. "The Evolution of War: The Fourth Generation." *Marine Corps Gazette* (September 1994).

_______. *The Sling and the Stone: On War in the 21st Century.* Minneapolis, MN: Zenith Press, 2006.

Harris, Brice F. *America, Technology and Strategic Culture: a Clausewitzian Assessment*. New York: Routledge, 2009.

Herodotus. *The Landmark Herodotus: The Histories*. Edited by Robert B. Strassler. Translated by Andrea L. Purvis. New York: Pantheon, 2007.

Holmes, Terence M. "Planning Versus Chaos in Clausewitz's On War." *The Journal of Strategic Studies* 30, no. 1 (February 2007): 129-51.

Howard, Michael, George J. Andreopoulos, and Mark R. Shulman, eds. *The Laws of War: Constraints on Warfare in the Western World*. New Haven, CT: Yale University Press, 1997.

Hyberg, Per. "Network-Centric Warfare and Information Theory." *Journal of Electronic Defense* 28, no. 12 (December 2005): 49-54.

Jasper, Scott, ed. *Securing Freedom in the Global Commons*. Stanford, CA: Stanford Security Studies, 2010.

Johnsen, William T., Douglas V. Johnson II, James O. Kievit, Jr., Douglas C. Lovelace, and Steven Metz. *The Principles of War in the 21st Century: Strategic Considerations*. Lexington, KY: CreateSpace, 1995.

Joint Chiefs of Staff. *Joint Pub 3-0: Joint Doctrine for Military Operations*. Washington, DC: U.S. Government Printing Office, 1995.

Jomini, Baron Antoine-Henri de. *The Art of War*. Princeton, N.J.: Bottom of the Hill Publishing, 2011.

Kramer, Franklin D., Stuart H. Starr, and Larry Wentz, eds. *Cyberpower and National Security*. New York: Potomac Books Inc., 2009.

Krapp, Peter. "Terror and Play, or What Was Hacktivism?" *Grey Room* 21 (Fall 2005): 70-93.

Kuhn, Thomas S. *The Structure of Scientific Revolutions,*. [2d ed. Vol. 2, of *International Encyclopedia of Unified Science. Foundations of the Unity of Science, V. 2, No. 2*. Chicago: University of Chicago Press, 1970.

Leonhard, Robert. *The Principles of War For the Information Age*. New York: Presidio Press, 2000.

Libicki, Martin C. *Cyberdeterrence and Cyberwar*. New York: Rand Publishing, 2009.

Lind, William S. *Maneuver Warfare Handbook*. Boulder, Colo.: Westview Press, 1985.

_______. "Understanding Fourth Generation War." Antiwar. Entry posted January 15, 2004. http://www.antiwar.com/lind/index.php?articleid=1702 (accessed January 3, 2012).

Lonsdale, David J. *The Nature of War in the Information Age: Clausewitzian Future (Strategy and History)*. New York: Routledge, 2004.

Lynn, William J. "Defending a New Domain." *Foreign Affairs* 89, no. 5 (Sep/Oct 2010): 97-108.

Mahan, Alfred Thayer. *The Influence of Sea Power Upon History. 1660-1783. [With Maps and Plans.]*. Charleston: British Library, Historical Print Editions, 2011.

McIvor, Anthony D., ed. *Rethinking the Principles of War*. Reprint ed. Annapolis, MD: US Naval Institute Press, 2007.

Meilinger, Phillip S., ed. *The Paths of Heaven: The Evolution of Airpower Theory*. Maxwell Airforce Base, AL: Air Univ Press, 1997.

Moffat, James. *Complexity Theory and Network Centric Warfare.* Washington, DC: Cforty Onesr Cooperative Research, 2003.

Osinga, Frans P.B. *Science, Strategy and War: The Strategic Theory of John Boyd (Strategy and History)*. New York, NY: Routledge, 2006.

Owens, William A. *Lifting the Fog of War*. Baltimore: The Johns Hopkins University Press, 2001.

_______, Kenneth W. Dam, Herbert S. Lin, and editors. *Technology, Policy, Law, and Ethics Regarding U.S. Acquisition and Use of Cyberattack Capabilities*. Washington, DC.: National Academies Press, 2009.

Papaj, Christopher J. "Clausewitz and 21st Century Warfare." Master's thesis, Army War College, 2008.

Paret, Peter, Gordon A. Craig, and Felix Gilbert, eds. *Makers of Modern Strategy: from Machiavelli to the Nuclear Age*. Princeton, N.J.: Princeton University Press, 1986.

Phister, Paul W., Dan Fayette, and Emily Krzysiak. "The 'CyberCraft' Concept." *Military Technology* 31, no. 9 (September 2007): 123-31.

Polk, Robert B. "A Critique of the Boyd Theory - Is It Relevant to the Army?" *Defense Analysis* 16, no. 3 (2000): 257-76.

Rattray, Greg. *Strategic Warfare in Cyberspace*. New York: The MIT Press, 2001.

Rice, Mason, Jonathan Butts, and Sujeet Shenoi. "A Signaling Framework to Deter Aggression in Cyberspace." *International Journal of Critical Infrastructure Protection* 4, no. 2 (August 2011): 57-65.

Schelling, Thomas C. *Arms and Influence*. 1966. Reprint, New Haven, CT.: Yale University Press, 2008.

Schrofl, Josef, Bahram M. Rajaee, and Dieter Muhr, eds. *Hybrid and Cyber War as Consequences of the Asymmetry: a Comprehensive*

Approach Answering Hybrid Actors and Activities in Cyberspace: Political Social and Military Responses. New York: Peter Lang Pub Inc, 2011.

Simpkin, Richard E. *Race to the Swift: Thoughts On 21st Century Warfare.* Revised ed. London: B.T. Batsford, 1998.

Smith, Edward A. Jr. "Network-Centric Warfare." *Naval War College Review* 54, no. 1 (Winter 2001): 59-61.

Smith, Thomas W. "The New Law of War: Legitimizing Hi-Tech and Infrastructural Violence." *International Studies Quarterly* 46, no. 3 (September 2002): 355-74.

Strachan, Hew, and Andreas Herberg-Rothe, eds. *Clausewitz in the Twenty-First Century*. New York: Oxford University Press, USA, 2007.

Streland, Arnold H. "Clausewitz on Space: Developing Military Space Theory through a Comparative Analysis." Master's thesis, Air Command and Staff College, Air University, 1999.

Sumida, Jon Tetsuro. *Inventing Grand Strategy and Teaching Command: The Classic Works of Alfred Thayer Mahan Reconsidered (Woodrow Wilson Center Press S.)*. Charleston: The Johns Hopkins University Press, 1999.

Vego, Milan N. *Operational Warfare: NWC 1004*. Newport, RI: U.S. Naval War College, 2000.

_______. "Systems Versus Classical Approach to Warfare." *Joint Force Quarterly* 52 (Winter 2009): 40-48.

Waldman, Thomas. "'Shadows of Uncertainty': Clausewitz's Timeless Analysis of Chance in War." *Defence Studies* 10, no. 3 (September 2010): 336-68.

Watts, Barry D. *Clausewitzian Friction and Future War (McNair Paper 52).* Washington, DC: National Defense University, 2004.

Woods, David D., and Erik Hollnagel. *Joint Cognitive Systems: Patterns in Cognitive Systems Engineering*. Boca Raton: CRC Press, 2006.

Zimm, Alan D. "Human-Centric Warfare." *U.S. Naval Institute Proceedings* 125, no. 5 (May 1999): 28-31.

About the Author

David Aucsmith is a Senior Principal Research Scientist at the Applied Physics Laboratory, University of Washington and an Affiliate Professor in Computer Science & Electrical Engineering at the University of Washington. Additionally, he is currently Chief Scientist at DarkLight AI, where he works on Ontological -Driven Artificial Intelligence Systems for cyber security.

Prior to his current positions, David Aucsmith was Senior Director of Microsoft's Institute for Advanced Technology in Governments. Before joining Microsoft, Aucsmith was the chief security architect for Intel Corporation.

Aucsmith is a former officer in the U.S. Navy and has been an industry representative to numerous international, government and academic organizations including the Georgia Institute of Technology School of Computer Science, the Software Engineering Institute at Carnegie Mellon University, and National Academies Board on Army Research. Aucsmith was also U.S. industry representative to the G8 Committee on Organized, Transnational, and Technological Crime where he participated directly in the G8 summits in Paris, Berlin and Tokyo.

He is the author of numerous papers and currently lectures at the Naval Postgraduate School. Additionally, he writes historical fiction novels.

Made in United States
North Haven, CT
04 November 2026

21765617R00035